Food Energy in Ecosystems

D1540793

Printed in Mexico

ISBN-13: 978-0-15-366679-7

ISBN-10: 0-15-366679-X

2 3 4 5 6 7 8 9 10 805 16 15 14 13 12 11 10 09 08

SCHOOL PUBLISHERS

Visit *The Learning Site!*
www.harcourtschool.com

What Is an Ecosystem?

VOCABULARY

environment
ecosystem
population
community

It is important to protect the **environment**. The environment is made up of living and nonliving things.

Changing the direction of a flowing river can hurt the river's **ecosystem**.

People sometimes build homes where prairie dogs live. This makes the space for the animals smaller. Then the **population** of prairie dogs must live with this change.

If a whole population of trees dies, the **community** of birds and animals that depended on them will suffer and change.

The **main idea** is what the text is mostly about.
Details tell more about he main idea.
Look for **details** that describe an ecosystem.

Ecosystems

All the things around you make up the **environment**.
An environment includes living things. Living things
are people, animals, and plants. Nonliving things are
in an environment, too. Nonliving things include water
and air. Soil and weather are nonliving too. Your room
environment has people, paper, light, air, and much more.

All the things that interact in an environment make up
an **ecosystem**. An ecosystem has two parts—living things
and nonliving things.

An ecosystem can be small or large, wet or dry, cold
or hot.

The parts of an ecosystem work
together. Dead plants make the soil
healthy. Soil and water help new plants
grow. Animals eat plants.

4

◀ Under a rock, you might find an ecosystem of worms, stones, and bits of wood.

An ecosystem can be under a rock. Wet soil, insects, and worms are in that ecosystem.

An ecosystem can be large. A forest is a large ecosystem. Trees, animals, and rocks are in a forest ecosystem.

An ecosystem has a *climate*. The climate can be hot, warm, or cold. The climate can be wet or dry. The desert has a dry climate. In a forest, the climate might be cool and damp. Changes in climate affect ecosystems.

 What two parts make up an environment?

In a forest ecosystem, the climate can be rainy or dry, cold or hot.

Individuals and Populations

One deer is an *individual*. A group of deer is called a **population**. A population is a group of the same kind of plant or animal living in the same ecosystem. A blue jay is an individual. A group of blue jays is a population. Robins are birds, too. But robins are a different population, because robins are not the same as blue jays.

The individual water lily is part of a population of water lilies.

◀ The red-winged blackbird can live in different ecosystems.

Some populations can live in different ecosystems. Some birds live in dry places. They can also live in wet places. If an ecosystem changes, the birds can fly somewhere else.

Some populations must live in only one kind of ecosystem. If the ecosystem changes, the population will have a hard time.

 Name a population you might see at a pond.

▲ This ecosystem is a cypress swamp. This ecosystem is named for the main population of trees that live there.

Communities

Plants, animals, and people live together. They live in a community. A **community** is all the populations that live in the same place.

The Blue Ridge Mountains area is a community. Many populations live in this community. There are different tree populations. There are different animal populations. There are insect and bird populations. All these populations live together in the Blue Ridge.

In the Blue Ridge, you will find many different populations.

There are different kinds of forests in Georgia. The Okefenokee Swamp has a forest community. The populations there are different from the Blue Ridge. For example, cypress trees grow in the swamp forest, while oak trees grow in the Blue Ridge forest.

Some populations, such as black bears, live in both communities.

 How can communities be different?

These sand dunes on Cumberland Island are part of a coastal community. ▶

Review

Complete this main idea statement.

1. All the living and nonliving things in an area form an _____ .

Complete these details statements.

2. An individual ladybug is part of a ladybug _____ .

3. Many different populations live in the same _____ .

4. The _____ includes all the living and nonliving things that surround you.

What Are the Roles of Producers, Consumers, and Decomposers?

VOCABULARY

producer
consumer
herbivore
carnivore
omnivore
decomposer

A plant is a **producer** because it can make its own food.

A ladybug is a **consumer** because it must eat other living things.

A cow is an **herbivore**. It eats only plants.

A bobcat is a **carnivore**. It eats other animals.

A bear eats both plants and animals. It is an **omnivore**.

Sow bugs eat dead plants. They are one kind of **decomposer**. Other decomposers break down parts of dead animals.

READING FOCUS SKILL
MAIN IDEA AND DETAILS

The **main idea** is what the text is mostly about. **Details** tell more about he main idea.

Look for **details** that describe how living things get energy from other living things.

Producers and Consumers

Green plants make their own food. Their leaves use the energy in sunlight to make food. Any living thing that makes its own food is called a **producer**. Plants are producers. Producers can be small, like moss. Producers can be large, like an oak tree.

▲ Plants use energy from sunlight to make food. Without the sun, plants would die.

A bobcat gets its energy from other consumers.

Some animals eat plants. The energy stored in the plants is used by the animal's body. Deer and cows eat plants.

Some animals eat other animals. Lions and hawks eat other animals. They do not eat plants.

An animal that eats plants or other animals is called a consumer. **Consumers** cannot make their own food the way plants can. Consumers must eat other living things. Consumers eat plants or other animals.

(Focus Skill) **Give one example of a producer and one example of a consumer.**

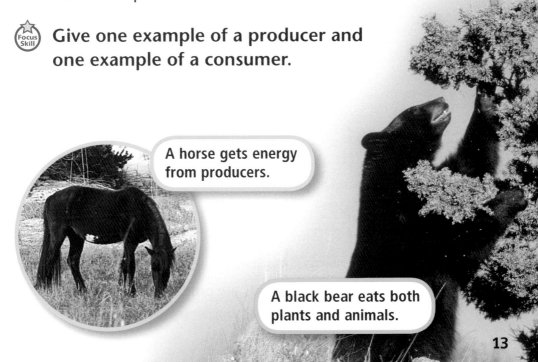

A horse gets energy from producers.

A black bear eats both plants and animals.

Kinds of Consumers

There are three kinds of consumers. A **herbivore** is an animal that eats only plants, or producers. Horses are herbivores. Squirrels and rabbits are herbivores. They also eat producers.

A **carnivore** is an animal. It eats only other animals. Lions are carnivores. A carnivore can be large, like a whale. A carnivore can be small, like a frog. Carnivores eat other consumers.

An **omnivore** is an animal. It eats both plants and animals. Bears are omnivores. So are many people. They eat producers and consumers.

 Give an example of a herbivore, a carnivore, and an omnivore.

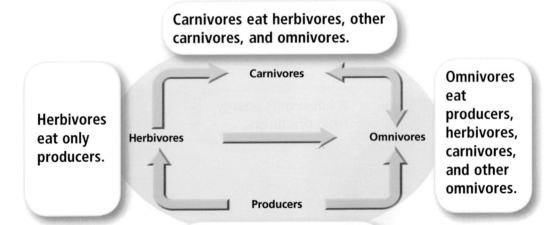

This diagram shows how consumers get energy to live. The arrows show the direction of energy flow.

Carnivores eat herbivores, other carnivores, and omnivores.

Herbivores eat only producers.

Omnivores eat producers, herbivores, carnivores, and other omnivores.

Carnivores

Herbivores

Omnivores

Producers

Green plants are producers.

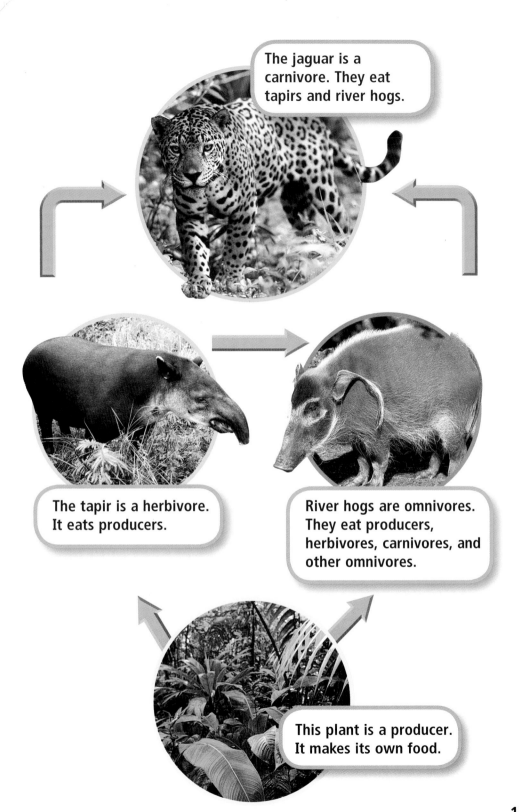

The jaguar is a carnivore. They eat tapirs and river hogs.

The tapir is a herbivore. It eats producers.

River hogs are omnivores. They eat producers, herbivores, carnivores, and other omnivores.

This plant is a producer. It makes its own food.

Decomposers

A decomposer is a living thing. A **decomposer** eats dead plants and animals. Decomposers also break down waste. A decomposer can make soil healthy. Good soil helps plants grow. Animals eat the healthy plants. This keeps the animals healthy, too.

This fungus is a decomposer. It grows on dead trees. The fungus helps break down the wood.

Earthworms are decomposers. They eat dead plants.

Fungi are decomposers, too. Fungi break down wood from dead trees.

Many bacteria are decomposers. Some bacteria live in the soil. Some bacteria live in water. Bacteria break down parts of dead plants and animals.

Decomposers are very helpful. Without them, dead plants and animals would cover Earth.

 Name three kinds of decomposers.

Bacteria can be found in water and soil. ▶

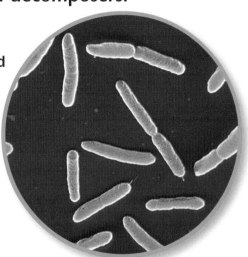

Review

Complete this main idea sentence.

1. Herbivores, carnivores, and omnivores get energy from other _____ things.

Complete these details statements.

2. A _____ is something that can make its own food.

3. A _____ must eat other living things.

4. A _____ feeds on dead plants and animals.

How Does Energy Flow Through an Ecosystem?

VOCABULARY

habitat
niche
food chain
prey
predator
food web
energy pyramid

A chipmunk's *habitat* is in the woods. Trees are all around. Acorns are on the ground. A **habitat** is an environment that meets the needs of a living thing.

Some kinds of turtles dig a hole in the sand. They lay eggs in the hole. They eats grasses. These are parts of the turtle's **niche**, or role, in its ecosystem.

Chipmunks are found in this owl's **food chain**, which shows the flow of energy among living things.

A hawk catches a fish. The fish is **prey**.

An owl is a **predator**. It eats mice and snakes.

Snakes eat mice. Hawks eat mice and snakes. These overlapping food chains make up a **food web**.

This **energy pyramid** shows how much energy moves from producers to consumers in a food chain.

Habitats

Plants and animals live in certain places. This is a habitat. A **habitat** is an environment that meets the needs of a living thing. Some plants and animals meet their needs in a desert. Some live in the sea. Others live where it is cold.

Different plants and animals can share a habitat. For example, spiders, snakes, and plants live together in the same desert habitat.

Sidewinders, tarantulas, and sagebrush all live in the same habitat. The desert habitat meets all their needs. ▶

Tarantula

Sagebrush

Sidewinder

This snake has a niche. Its niche helps to balance the number of small animals in its desert habitat.

Each living thing has a role, or **niche**. A niche is how a living thing works within the habitat. A niche includes these things:

- where a plant or animal lives.
- how it reproduces.
- where an animal gets food.
- how an animal stays safe.

A sidewinder is a snake. It has venom. The venom helps the snake kill mice and birds. Using venom to kill is part of the snake's niche.

If these snakes died, the desert would have too many mice and birds. There is not enough food in the desert for all these animals. They would begin to starve. The snake's niche keeps the number of small animals in balance.

 Describe a sidewinder's habitat and niche.

Food Chains

Plants and animals depend on one another. Food energy moves from one living thing to another. A **food chain** shows the path of food energy. Every food chain starts with producers, or plants.

An oak tree is a producer. Acorns grow on oak trees. Chipmunks eat acorns. The chipmunk is a consumer. Then a hawk eats the chipmunk. The hawk is a consumer, too.

This is how food energy moves. It starts with the acorn. It moves to the chipmunk. Then it moves to the hawk. This is a food chain.

Consumers that are eaten are called **prey**. Consumers that eat prey are called **predators**. The chipmunks are prey. Hawks are predators. Owls are predators, too. Owls eat chipmunks. Predators, such as hawks and owls, compete for prey.

 Does a food chain start with a producer or a consumer?

Acorns provide energy for the chipmunk.

Then the hawk eats the chipmunk.

The chipmunk provides energy for the hawk.

Food Webs

A food chain shows how an animal gets energy. Different food chains can cross. For example, a hawk eats mice *and* small birds. When food chains cross, they make a **food web**. Food webs are on land. Food webs are in water, too.

Trace the different food chains, or paths, that make up this food web. For example, energy moves from plants to the mouse to the hawk. Energy also moves from plants to the snail to the sparrow to the hawk.

At the bottom of the food web are *first-level consumers*. Snails and insects are first-level consumers. Anything that eats a first-level consumer is called a *second-level consumer.* Snakes and birds are second-level consumers. They are eaten by the *top-level consumers.* Hawks and wolves are top-level consumers. Like all living things, when top-level consumers die, decomposers help break down their remains.

What happens after a first-level consumer eats a plant?

Energy Pyramids

In a food chain, energy is passed on. Energy moves from one thing to another. An **energy pyramid** shows *how much* energy is passed on.

Producers are at the bottom of the energy pyramid. Plants get energy from the sun. Plants use about 90 percent of all this energy. They use the energy to grow. The extra 10 percent is stored. Plants store extra energy in their roots, stems, and leaves.

A wolf must eat many smaller animals to survive.

There are fewer animals toward the top of an energy pyramid.

Then plants are eaten by squirrels and rabbits. These small consumers use up 90 percent of the energy from the plants. The other 10 percent of the energy is stored in their bodies.

The smaller animals are eaten by owls and foxes. These larger consumers use up 90 percent of the energy from the smaller animals. The other 10 percent is stored in their bodies.

Small amounts of energy are passed on. Energy moves from one living thing to the next. Each level passes a little energy to the next. The first-level consumers need many producers. That is why the bottom of the energy pyramid is wide.

 What starts the flow of energy at the bottom of an energy pyramid?

Review

Complete these sequence statements.

1. A snake's _____ helps to make a balanced habitat.

2. Every _____ chain starts with producers.

3. First-level consumers are eaten by second-level _____ .

4. An energy _____ ends with top-level consumers.

What Factors Influence Ecosystems?

Insects and plants need each other. They are **biotic**, or living, parts of the ecosystem.

Rocks and water are not alive. They are nonliving. They are the **abiotic** parts of the ecosystem.

A cause is something that makes another thing happen. An effect is the thing that happens.

Look for examples of cause and effect when you read about climate in an ecosystem.

Living Things Affect Ecosystems

Plants are the living parts of an ecosystem. Animals are living parts of an ecosystem, too. Living parts of an ecosystem are **biotic**. *Bio* means "life."

Plants affect animals in many ways. Plants are food for animals and insects. If a plant dies, there is less food. Plants, like trees, are also homes for animals and insects.

Animals affect plants in many ways. Birds and deer spread seeds. This causes plants to grow in new places. Animal droppings make the soil healthy. This helps the plants to grow. If too many animals eat a plant, they can kill it.

Plants and animals are biotic. They can change an ecosystem.

 How can an insect change an ecosystem?

◀ Hungry moth caterpillars eat leaves on a tree.

A healthy tree is not hurt when a few caterpillars nibble on the leaves.

Too many caterpillars can eat all the leaves on a tree. This makes the tree weak.

Nonliving Things Affect Ecosystems

Plants and animals are living parts of an ecosystem. Ecosystems have nonliving parts, too. Sunlight, air, water, and soil are nonliving. These nonliving parts are **abiotic**. They are not alive. But they can change an ecosystem.

Look at the picture on these pages. The sun's light is shining. Plants need sunlight to grow and make food. Sunlight is an abiotic factor.

Water is abiotic. Plants need water to grow. Fish must live in water. Animals need to drink water. Water affects the ecosystem.

Soil is abiotic. Many plants need soil to grow. Some soil is dry and hard. Some soil is soft and wet. Soil affects the kinds of plants that can grow in an ecosystem.

 What will happen to an ecosystem if a river dries up?

Living parts of an ecosystem need the nonliving parts.

Climate Affects Ecosystems

What is the climate where you live? Is it hot, cold, or warm? Do you have a lot of rain, or is it dry where you live? Do you have a lot of sunlight? Climate is an abiotic factor. Climate is the amount of

- sunlight that shines in a place.
- rain that falls in a place.
- warm air that is in a place.
- cold air that is in a place.

Climate affects the soil in an ecosystem. Climate affects the plants that grow in an ecosystem. Climate also affects the animals that live in an ecosystem.

 How does a dry climate affect an ecosystem?

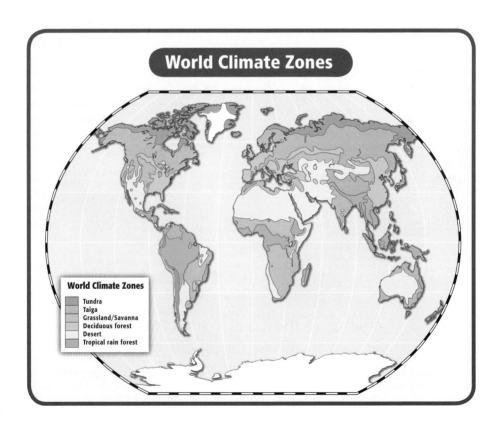

World Climate Zones

World Climate Zones
- Tundra
- Taiga
- Grassland/Savanna
- Deciduous forest
- Desert
- Tropical rain forest

◀ Some climates have four seasons.

A rain forest has a wet climate. ▶

◀ A desert has a dry climate.

Review

Focus Skill **Complete these cause and effect statements.**

1. _____ factors, like birds spreading seeds, affect the ecosystem.

2. _____ factors, like sunlight, affect the ecosystem.

3. A change in the _____ , such as no rain falling, can cause a change in an ecosystem.

4. Living and _____ things affect an ecosystem.

GLOSSARY

abiotic (ay•by•AHT•ik) Describes a nonliving part of an ecosystem

biotic (by•AHT•ik) Describes a living part of an ecosystem

carnivore (KAHR•nuh•vawr) An animal that eats only other animals

community (kuh•MYOO•nuh•tee) All the populations of organisms living together in an environment

consumer (kuhn•SOOM•er) A living thing that can't make its own food and must eat other living things

decomposer (dee•kuhm•POHZ•er) A living thing that feeds on the wastes of plants and animals

ecosystem (EE•koh•sis•tuhm) A community and its physical environment together

energy pyramid (EN•er•jee PIR•uh•mid) A diagram showing how much energy is passed from one organism to the next in a food chain

environment (en•VY•ruhn•muhnt) All of the living and nonliving things surrounding an organism

food chain (FOOD CHAYN) A series of organisms that depend on one another for food

food web (FOOD WEB) A group of food chains that overlap

habitat (HAB•ih•tat) An environment that meets the needs of an organism

herbivore (HER•buh•vawr) An animal that eats only plants, or producers

niche (NICH) The role of an organism in its habitat

omnivore (AHM•nih•vawr) An animal that eats both plants and other animals

population (pahp•yuh•LAY•shuhn) All the individuals of the same kind living in the same ecosystem

predator (PRED•uh•ter) A consumer that eats prey

prey (PRAY) Consumers that are eaten by predators

producer (pruh•DOOS•er) A living thing, such as a plant, that can make its own food